THE GALACTIC VOID:

UNDERSTANDING COSMIC VOIDS

AND INTERGALACTIC SPACE

D.A. Nyberg, J.D., Ph.D.

Cover designed with DALL-E and Canva

DEDICATION

This little book is dedicated to those who have

held a Yale Bright Star Catalog

and a BD (Bonner Durchmusterung)

and know which one(s) do not have

constellation designations.

FORWARD

In the mid-1980's, after completing my dissertation I latched onto my interest in astronomy for a time. An astronomer in Hawaii, I'll not give her name, encouraged me to find a niche and join the labor in astronomy. Soon my reading drew my interest to exploring "the" galactic void. My mentor encouraged me to submit a request to the Very Large Array. She sent me the manual and I set to work crafting a proposal.

When the proposal was complete, I sought review fearful that I would submit a foolish and unworthy proposal. One person encouraged me to submit the proposal and two others discouraged me.

At about the same time I was admitted to law school, then a decade and a half of litigation and then onto Advanced Studies under Boston College's Harvey D. Egan followed by work in behavioral health/substance use disorder treatment.

Now that I have retired for the second time, I am indulging my interest in, now that I know better, galactic voids.

I am thankful for the time to research the subject and to share what I have found.

I hope you enjoy the text and please let me know of my errors.

D. A. Nyberg, J.D., Ph.D.

TABLE OF CONTENTS

Chapter 1

INTRODUCTION TO GALACTIC VOIDS

Overview

Galactic voids are among the most fascinating features of the cosmos, representing vast regions of space with remarkably low densities of matter. Unlike clusters or filaments, which are densely populated with galaxies and other celestial structures, voids are sparsely populated, often extending for tens to hundreds of millions of light-years with only a handful of galaxies dispersed within them. These voids make up roughly 80% of the universe's volume, yet they contain less than 20% of its total mass, primarily in the form of dark matter and sparse hydrogen gas.

The study of voids offers valuable insights into the large-scale structure of the universe. Within the context of the cosmic web—the grand network of filaments, sheets, and voids that defines the universe's structure—voids are essential for understanding the distribution and behavior of both visible and dark matter. Void dynamics are shaped by cosmic expansion, allowing researchers to test cosmological models and study the effects of dark energy, the mysterious force driving the accelerated expansion of the universe. Additionally, voids are unique environments where galaxy formation is rare, thus allowing scientists to observe the gravitational effects and structural behaviors of dark matter without significant interference from luminous matter, providing a clearer view into the universe's underlying structure.

Galactic Voids in the Universe

The discovery of galactic voids was a pivotal moment in our understanding of cosmic structure, radically altering the perception of the universe's distribution of matter. In the early 1970s, astronomers began conducting redshift surveys— measurements of the distances to galaxies based on their spectral redshifts, which indicate how fast a galaxy is moving away from us due to the expansion of the universe. These early surveys, conducted with optical telescopes, revealed a non-uniform distribution of galaxies, with numerous dense regions interspersed with large, seemingly empty spaces.

One of the first major discoveries of a galactic void was the Bootes Void, identified in 1981 by astronomers Robert Kirshner and colleagues. The Bootes Void spans approximately 330 million light-years and contains only a few dozen galaxies, where, by comparison, a similarly sized region in a dense part of the cosmic web might contain thousands. This discovery confirmed the existence of significant voids and hinted at a much larger structure in the cosmos. Further redshift surveys, particularly those conducted by the Sloan Digital Sky Survey (SDSS), revealed that galaxies are organized into vast interconnected networks with voids separating the clusters and filaments.

Voids are not only vast in scale but also central to understanding the cosmic web's structure. Their boundaries, known as void walls, consist of galaxy clusters and filaments, which create the scaffold-like structure of the universe. Observations show that voids play a critical role in the evolution of this large-scale structure. As the universe expands, these voids grow larger, pushing galaxies and clusters closer together along the edges of the voids. This growth and the movements of galaxies along void walls are important for studying cosmic flows—the large-scale movements of matter within the universe. In studying these phenomena, scientists gain a better understanding of dark energy, as the rate at which voids expand is sensitive to the presence of this energy.

Historical Context

The exploration of galactic voids has been closely tied to advancements in observational astronomy, particularly in large-scale sky surveys and radio observations. During the 1970s and 1980s, astronomers embarked on several large-scale galaxy redshift surveys, mapping the universe in three dimensions. Early surveys were limited in scope, yet they provided the first hints of voids by revealing large gaps in galaxy distribution. These gaps puzzled astronomers and challenged existing theories that assumed a more uniform distribution of galaxies.

The Very Large Array (VLA), a powerful radio telescope in New Mexico, was instrumental in advancing void studies. By detecting faint hydrogen gas emissions in and around voids, the VLA allowed astronomers to map low-density regions that are not visible in optical wavelengths. These radio observations confirmed that voids were not entirely empty but contained faint clouds of neutral hydrogen, which are crucial for studying the gravitational dynamics and distribution of dark matter in these regions. The VLA's high sensitivity to radio wavelengths enabled researchers to trace these emissions across vast distances, providing key evidence of the large-scale cosmic web structure and the behavior of void boundaries.

Space-based telescopes further revolutionized void research. The Hubble Space Telescope (HST), with its capacity to capture deep-field images, identified faint galaxies near void peripheries, reinforcing the concept that voids were not completely devoid of matter. These observations revealed isolated galaxies and groups residing in low-density environments, providing a natural laboratory for studying galaxy evolution under minimal gravitational influence from other galaxies. The Hubble Deep Field observations in the 1990s were particularly insightful, allowing astronomers to study galaxy properties in the early universe, including those located near void boundaries.

The Sloan Digital Sky Survey (SDSS), initiated in 2000, provided an unprecedented view of the universe by mapping the positions, distances, and velocities of millions of galaxies. SDSS data clearly illustrated the large-scale structure of the universe, with voids interwoven among dense galaxy filaments. With the SDSS, astronomers could quantify the shape, size, and spatial distribution of voids, greatly enhancing the precision of void research. SDSS data allowed cosmologists to construct detailed models of the universe's structure, highlighting voids as key features that play a major role in cosmic evolution.

Subsequent observations by the Planck satellite and other cosmic microwave background (CMB) observatories further contributed to void research by studying the effects of voids on CMB photons. As photons pass through these low-density regions, they are slightly altered in a process known as the Integrated Sachs-Wolfe effect. Observations of this effect offer indirect evidence of dark energy and help refine measurements of the universe's expansion rate, which is influenced by the distribution of voids and other large-scale structures. Through these contributions, the Planck satellite provided a critical layer of data, connecting the study of voids with fundamental questions about the universe's origins and expansion.

The study of voids remains a dynamic area of research, with ongoing surveys and forthcoming projects poised to deepen our understanding. Observatories like the Vera C. Rubin Observatory and the Square Kilometre Array (SKA) will provide astronomers with even more powerful tools to probe voids. These projects will map the distribution of galaxies and hydrogen gas across vast volumes, potentially revealing new details about void structure and growth over cosmic time. By studying voids with these cutting-edge instruments, scientists continue to unravel the mysteries of the universe's large-scale structure, dark energy, and the fundamental processes shaping cosmic evolution.

Chapter 2

OBSERVING GALACTIC VOIDS

Telescopic Techniques

Detecting and studying galactic voids requires sophisticated telescopic methods that can observe large regions of space and detect faint signals across vast distances. Since voids are characterized by a scarcity of luminous matter, they are typically identified through redshift surveys and density mapping, which reveal the large, empty regions between galaxy clusters and filaments. Two primary types of telescopes are instrumental in this research: radio telescopes and optical/infrared telescopes.

Radio Telescopes: The Very Large Array (VLA) in New Mexico is one of the foremost radio observatories used to study voids. Radio telescopes are advantageous for detecting faint emissions of neutral hydrogen gas (HI) within voids. Since voids lack the concentrated galaxies and stars typically observed in optical wavelengths, radio wavelengths offer a clearer view of the sparse matter distributed throughout these regions. By detecting the 21-cm emission line of hydrogen gas, the VLA provides astronomers with a map of low-density regions, which helps outline void boundaries and examine gas distributions. This mapping technique is particularly useful in distinguishing void interiors from the denser filaments surrounding them. The VLA's sensitivity to large-scale structure allows it to trace faint signals across considerable distances, helping researchers to chart the dimensions and density profiles of some of the largest known voids.

Optical and Infrared Telescopes: Ground-based optical telescopes, such as those at the Keck Observatory, are also used in void research to perform redshift surveys that map galaxy positions and velocities. By measuring the redshifts of galaxies, astronomers can calculate distances, thereby revealing the 3D distribution of galaxies and the empty regions between them. However, optical telescopes have limitations in void studies, as the lack of bright galaxies in void regions reduces the amount of visible light available for observation. This is where infrared capabilities, such as those of the Spitzer Space Telescope, become advantageous, as they allow astronomers to detect cooler, fainter objects near void edges. These detections help confirm that voids are not entirely devoid of matter, though their densities are minimal compared to more populated regions.

The Role of Space-Based Observatories

Space-based telescopes play an essential role in observing galactic voids by providing high-resolution images and access to a wide range of wavelengths that are difficult or impossible to observe from Earth. Space telescopes avoid atmospheric interference, which can distort or obscure faint signals, allowing for deeper and more precise observations of void regions.

Hubble Space Telescope (HST): The Hubble Space Telescope has been instrumental in the study of voids, particularly through its deep-field observations, which capture faint galaxies along void peripheries. The Hubble Deep Field and subsequent ultra-deep field surveys revealed the distribution of galaxies across vast cosmic distances, helping astronomers understand the boundaries and shapes of nearby voids. These observations also highlighted the scarcity of galaxies within voids, reinforcing their status as low-density regions and providing visual confirmation of the cosmic web's filamentary structure.

Spitzer Space Telescope: Observing in the infrared spectrum, the Spitzer Space Telescope has contributed to void studies by detecting faint, cool objects near the edges of voids. Spitzer's sensitivity to infrared light enables the detection of older, dust-enshrouded galaxies that may reside in these low-density environments. Spitzer's observations support the notion that voids are not entirely empty but contain isolated, faint galaxies, offering insights into how galaxy evolution occurs in low-density environments. By comparing data from Spitzer and Hubble, astronomers have developed a more comprehensive understanding of the faint galaxy population associated with voids.

James Webb Space Telescope (JWST): The recently launched James Webb Space Telescope (JWST) promises to deepen our understanding of galactic voids with its advanced infrared capabilities, allowing t to observe extremely distant and faint objects with unparalleled precision. JWST's ability to detect faint light from the earliest galaxies will help astronomers study the population of galaxies near and within voids, providing data on the formation and evolution of voids over cosmic time. By observing the faintest objects at the edge of observable space, JWST may reveal how early voids formed and evolved, offering new insights into the structure of the early universe.

Challenges in Void Observation

Observing galactic voids presents several challenges due to their unique characteristics. Unlike galaxy clusters or filaments, voids are defined by their lack of luminous matter, making them inherently difficult to detect. The main challenges include the absence of bright light sources, the vast distances involved, and the need for high-sensitivity instruments to detect low-density matter.

Absence of Light Sources: Since voids contain very few galaxies, they emit minimal light. The lack of luminous matter complicates direct observation, requiring astronomers to rely on indirect methods such as redshift surveys and radio emissions to detect their presence. Redshift surveys allow astronomers to chart galaxy positions and identify voids by locating regions with few or no galaxies, but this method requires extensive surveys of vast areas of the sky.

Vast Distances and Low-Density Matter: Galactic voids often span hundreds of millions of light-years, making observations challenging due to the immense distances. At such scales, even faint signals from matter within voids become difficult to detect. The VLA's ability to pick up faint radio emissions is a significant advantage, yet the low-density nature of void matter means that even with radio telescopes, signals are faint and require long observation times to accumulate sufficient data for analysis. The large size of voids also means that mapping them requires substantial amounts of telescope time and computing resources.

Technological and Sensitivity Limitations: Despite the capabilities of the VLA, Hubble, Spitzer, and JWST, observing voids continues to push the limits of astronomical technology. For example, the VLA, though powerful, has limitations in resolving fine-scale structures within voids due to the weak radio emissions from sparse hydrogen gas in these regions. Similarly, while the Hubble Space Telescope provides high-resolution optical images, it is limited to observing relatively nearby voids due to its sensitivity constraints. Future projects, such as the Square Kilometre Array (SKA) and the Vera C. Rubin Observatory, are expected to address some of these limitations by offering increased sensitvity and resolution, thereby enhancing our ability to study these elusive regions more effectively.

By combining the strengths of both ground-based and space-based observatories, astronomers continue to refine their understanding of galactic voids. Although challenges remain, the study of voids is advancing with each technological improvement, providing deeper insights into the distribution of matter, the forces shaping the cosmos, and the role of dark energy in cosmic evolution.

Chapter 3

THE STRUCTURE AND SCALE OF THE UNIVERSE

Cosmic Web and Void Formation

The large-scale structure of the universe, often referred to as the cosmic web, is a complex network of interconnected galaxies, clusters, filaments, and vast, empty regions known as voids. The cosmic web reveals a universe that is anything but homogeneous; instead, matter is distributed in a web-like pattern where dense nodes of galaxies and galaxy clusters connect through filamentary structures, with vast voids lying in between. This intricate pattern emerged from small fluctuations in the early universe, as matter and dark matter clumped together under gravity over billions of years. Observations of the cosmic microwave background (CMB), such as those from the Planck satellite, show these initial fluctuations, which later evolved into the massive structures we observe today.

Voids are integral to the cosmic web, as they provide the expansive spaces that allow galaxy clusters and filaments to form and grow along the web. Early studies revealed that these voids are not empty but are instead filled with low-density hydrogen gas, dark matter, and occasional isolated galaxies. The Very Large Array (VLA) played a critical role in mapping this cosmic web by detecting faint emissions of neutral hydrogen within and around voids. These emissions help outline the boundaries of voids and the filaments that intersect them, providing a three-dimensional view of the cosmic web's structure.

Superclusters, Filaments, and Sheets

Within the cosmic web, matter is primarily concentrated along filaments, sheets, and superclusters of galaxies. Filaments are long, thin chains of galaxies and galaxy clusters, stretching across tens to hundreds of millions of light-years. They act as the backbone of the cosmic web, connecting superclusters and threading through voids. These filaments are where most galaxies are located, forming dense highways of matter. The Sloan Digital Sky Survey (SDSS), one of the most comprehensive surveys of galaxies, mapped these structures in detail, revealing a universe structured like an intricate lattice of galaxy chains.

Superclusters are massive conglomerations of galaxy clusters and groups linked together along these filaments. The Virgo Supercluster, for example, is a dense region containing our Local Group (the cluster that includes the Milky Way) and thousands of other galaxies. Superclusters are bound by gravity but do not form gravitationally bound structures on the scale of clusters. Instead, they are linked by the cosmic web's network of filaments. Sheets are another large-scale feature in the cosmic web, representing thin, flat structures of galaxies and clusters that, like filaments, delineate the borders of voids. Sheets and filaments often intersect, forming nodes of extreme density where superclusters reside. Observational data from the SDSS and the VLA have mapped these intersections and revealed the three-dimensional complexity of the cosmic web.

These structural elements—superclusters, filaments, and sheets—surround voids, effectively creating the boundaries that define these empty regions. The gravitational pull of the denser filamentary regions acts to draw matter away from voids, resulting in low-density regions where galaxy formation is rare. This separation process means that, over billions of years, voids have expanded as filaments and clusters continue to draw galaxies into denser regions, leaving behind the vast empty spaces that characterize the cosmic web.

Scale of Voids

Galactic voids vary significantly in size, from small voids spanning a few million light-years to supervoids that extend over hundreds of millions of light-years. Typical voids measure around 20–50 million light-years across, although some can be much larger. For instance, the Bootes Void, one of the largest observed voids, spans roughly 330 million light-years and contains only a few dozen galaxies, whereas a region of similar volume within a dense filament could contain thousands.

The scale of voids depends largely on the initial density fluctuations in the early universe and the gravitational influence of surrounding structures. Smaller voids often form within larger voids, creating a hierarchical structure where voids are nested within one another. This hierarchical nature is evident in observations from the SDSS, which have mapped not only individual voids but also the cistribution of voids of varying sizes within the cosm c web. Supervoids, the largest class of voids, are relatively rare but have a profound influence on the dynamics of the cosmic web. They represent regions where gravitational forces from surrounding superclusters and filaments have effectively evacuated the area, leaving only a sparse population of galaxies and dark matter.

The study of void scales and their properties offers insights into the overall matter distribution in the universe, including the behavior of dark energy, which appears to accelerate the expansion of these low-density regions. Voids expand over time as galaxies are drawn toward denser regions of the cosmic web, a process observed and confirmed by both the SDSS and VLA. The expansion rate of voids offers indirect measurements of the effects of dark energy, making voids an important area of study for cosmologists interested in the fate of the universe.

Voids, superclusters, filaments, and sheets together form the cosmic web, defining the large-scale structure of the universe. This structure, as mapped by the SDSS and probed by radio telescopes like the VLA, reveals a universe of remarkable complexity. The relationship between voids and the denser regions that surround them continues to be a key focus in cosmology, helping astronomers understand the interplay between visible matter, dark matter, and dark energy across the universe.

Chapter 4

THE NATURE AND PROPERTIES OF

GALACTIC VOIDS

Physical Characteristics

Galactic voids are defined by their exceptionally low densities, often containing less than 20% of the average matter density found in denser regions of the universe, such as galaxy clusters and filaments. These voids typically extend over vast distances, from tens to hundreds of millions of light-years across. Within these regions, the lack of galaxies and the minimal presence of luminous matter create some of the emptiest and coldest spaces in the cosmos.

Density measurements in voids reveal that they primarily contain dark matter, sparse amounts of neutral hydrogen gas, and isolated or faint dwarf galaxies. Observations from the Very Large Array (VLA) have provided critical insights into the distribution of neutral hydrogen in voids, detecting faint emissions even in low-density regions. These hydrogen gas detections indicate that voids are not entirely empty, though the matter within them is spread over such large distances that it remains difficult to detect with optical instruments alone.

Temperature within voids is also notably lower than in denser cosmic structures. Because voids lack the frequent gravitational interactions that generate heat in galaxy clusters, their average temperatures are lower, allowing gases to exist in a cooler state. This low temperature, combined with sparse gas density, means that voids emit minimal detectable radiation, further complicating direct observations. Instruments like the Chandra X-ray Observatory have been used to study the temperature of void gas, although the low emission makes precise measurements challenging. Chandra has, however, detected occasional X-ray sources near void edges, indicating some interactions between void gas and adjacent galaxy clusters.

Dark Matter and Dark Energy in Voids

Dark matter and dark energy are fundamental components in understanding the behavior and evolution of voids. Although voids contain very little visible matter, they are influenced by the gravitational pull of dark matter, which is thought to account for most of the mass within them. Dark matter is believed to form a diffuse scaffold within voids, with gravitational effects that contribute to the void structure. Observations using the Chandra X-ray Observatory and the VLA have detected the effects of dark matter indirectly, as the distribution of hydrogen gas and isolated galaxies within voids aligns with where dark matter is expected to be concentrated.

Dark energy, the force driving the accelerated expansion of the universe, plays an equally crucial role in void dynamics. Since voids are low-density regions with weaker gravitational fields than galaxy clusters, they expand at a faster rate than denser regions. This accelerated expansion leads to the gradual enlargement of voids, pushing matter toward the denser regions that surround them and causing voids to grow over cosmic time. This growth, accelerated by dark energy, effectively evacuates voids, causing them to expand even as galaxies move toward the filaments and clusters on void boundaries. This expansion is particularly significant in larger voids, known as supervoids, which have a greater influence on the overall distribution of matter in the cosmic web.

The study of voids has provided indirect evidence of dark energy's role in cosmic expansion. By observing the rate at which voids expand and comparing this with theoretical models, astronomers have been able to measure dark energy's influence across vast cosmic scales. Voids offer a unique environment for these studies, as they allow researchers to observe cosmic expansion in regions where gravitational influences from visible matter are minimized. This makes voids essential for refining models of dark energy and for exploring its potential effects on the universe's long-term evolution.

Voids and Low-Temperature Gases

Although voids are defined by their low density, they do contain small amounts of neutral hydrogen gas and other intergalactic medium (IGM) elements. The VLA has detected faint 21-cm hydrogen emissions within some voids, allowing researchers to map the distribution of this sparse gas. These findings reveal that even in the universe's emptiest regions, neutral hydrogen remains present, albeit in extremely low concentrations. This gas is cold, with temperatures that often reach only a few degrees above absolute zero, as the lack of frequent gravitational interactions means that gases in voids remain largely unheated.

The presence of this cold, low-density gas has significant implications for cosmic evolution. First, it suggests that the intergalactic medium extends through voids, albeit in a highly diffused form. Second, the cold temperatures in void regions affect gas dynamics, as cooler gases do not collapse easily to form stars or galaxies. This low density and temperature of gas in voids thus prevent significant star formation, leading to the observed scarcity of luminous objects in these regions. Observations from the VLA confirm this pattern, as void interiors generally contain only a few dwarf galaxies, while most galaxies cluster around void peripheries where gas density and temperature are higher.

Interestingly, research on void gas may also provide insights into the early stages of cosmic evolution. The existence of this sparse hydrogen gas could offer clues about the primordial conditions of the universe shortly after the Big Bang, as it has remained relatively unaffected by the gravitational forces that have shaped galaxies and clusters. Studies of void gas distribution and composition may therefore help astronomers understand how matter behaved in the early universe before it coalesced into stars and galaxies.

In summary, galactic voids represent one of the most intriguing areas of cosmology, where the interaction of dark matter, dark energy, and low-density gas contributes to a unique environment largely devoid of luminous structures. The continued study of these regions, aided by observations from instruments like the Chandra X-ray Observatory and the VLA, offers insights into the distribution of dark matter, the effects of dark energy, and the conditions that shaped the early universe.

Chapter 5

THE ROLE OF VOIDS IN COSMIC EVOLUTION

Cosmic Expansion and Voids

Galactic voids provide a unique perspective on the nature of cosmic expansion. Unlike dense regions such as galaxy clusters where gravitational forces act to pull matter inward, voids represent low-density regions where gravity's influence is minimal, allowing the effects of cosmic expansion to become more pronounced. The uniform expansion of the universe, as initially described by Edwin Hubble, is evident in the way voids stretch and grow over time, pushing surrounding galaxies and matter into denser regions of the cosmic web. Voids essentially mirror the accelerating expansion of the universe on a large scale, serving as "cosmic laboratories" that help scientists study dark energy—the mysterious force driving this expansion.

The relationship between voids and cosmic expansion is further illustrated by the movement of galaxies away from void centers. The low-density environment within voids amplifies the effects of dark energy, making them expand faster than other structures. As a result, matter near void boundaries tends to migrate outward, moving toward the filaments and clusters that form the dense edges of voids. Observations from the Sloan Digital Sky Survey (SDSS) and the Very Large Array (VLA) have helped track these cosmic flows, revealing how galaxies appear to be pulled into these dense structures, further "evacuating" the void interiors.

Voids are instrumental in measuring the effects of cosmic expansion across vast distances. By observing the rate of expansion within these low-density regions, scientists can better understand the influence of dark energy on the universe's large-scale structure. Because voids are free from significant gravitational interference, they provide a clear view of the expansion process, making them ideal for testing theoretical models of dark energy and understanding its long-term impact on the cosmos.

Evolution of Voids Over Time

Voids are not static structures; they expand and evolve over cosmic time. The initia formation of voids is thought to have begun shortly after the Big Bang when small fluctuations in the density of matter set in motion a process of clumping and separaticn. Over billions of years, gravitational forces drew matter toward denser regions, creating the filaments and clusters that make up the cosmic web. Voids emerged as the leftover spaces, where the absence of strong gravitational forces allowed them to expand as surrounding matter moved away.

As time progresses, voids continue to grow larger as galaxies drift outward and concentrate along void boundaries. This process of void expansion, driven in part by dark energy, causes voids to merge and form larger structures known as supervoids. Observations from the Sloan Digital Sky Survey confirm that voids exhibit a hierarchical structure, with smaller voids often nested within larger ones. This structure allows voids to coalesce over time, creating vast empty regions that influence the dynamics of the surrounding cosmic web.

The expansion of voids also affects the distribution of galaxies in their vicinity. Studies from ground-based telescopes like the Vera C. Rubin Observatory have shown that galaxies near void boundaries exhibit outward velocity patterns, indicative of the "push" effect caused by void expansion. This movement causes galaxies to cluster more tightly along filamentary regions and supercluster nodes, intensifying the contrast between dense and sparse regions in the cosmic web. Thus, as voids grow, they help shape the large-scale distribution of matter in the universe, reinforcing the cosmic web's structure and defining its evolution.

Voids and Galaxy Formation

One of the most intriguing aspects of voids is their lack of galaxy formation. Voids are characterized by minimal star formation, sparse matter distribution, and a general absence of large galaxies. Theories explaining this phenomenon suggest that the low density and temperature of gases within voids prevent the necessary gravitational collapse for galaxy formation.

Without sufficient matter to generate gravitational clumping, stars and galaxies cannot form at the same rates seen in denser regions.

The limited presence of matter in voids has significant implications for galaxy evolution. Studies from the Hubble Space Telescope and Spitzer Space Telescope indicate that the few galaxies found within voids tend to be isolated and relatively small, often classified as dwarf galaxies. These galaxies are also older and have less active star formation, suggesting that they formed at an earlier time in denser regions before drifting into the voids. The lack of new galaxies forming in voids reinforces the idea that galaxy evolution is heavily influenced by environment; dense regions foster ongoing galaxy formation, while voids suppress it.

Theories about galaxy formation in voids are further supported by the findings from the VLA, which has detected low-density hydrogen gas within voids but in insufficient quantities to trigger significant star formation. The cold temperatures of this gas also limit its collapse into galaxy-forming structures. This absence of gravitational collapse within voids highlights the role of environment in galaxy formation and suggests that voids serve as natural boundaries for galactic evolution. Observing the rare galaxies within voids provides a unique look into the effects of isolation on galactic properties, offering insights into the ways in which galaxies interact with and adapt to their surroundings over cosmic time.

In conclusion, galactic voids play a crucial role in cosmic evolution. Their structure and growth reflect the universe's expansion, particularly the effects of dark energy, while their influence on galaxy formation emphasizes the importance of environment in cosmic development. Through continued observations from ground-based and space-based surveys, scientists are gaining a deeper understanding of how voids contribute to the large-scale organization and evolution of the universe.

Chapter 6

HOW VOIDS ARE DETECTED

Techniques in Void Detection

Detecting galactic voids requires specialized methodologies that can reveal the large-scale distribution of matter in the universe. Voids, by definition, lack luminous matter, making them difficult to observe directly. Instead, astronomers rely on indirect techniques, such as redshift surveys, the Doppler effect, and gravitational lensing, to identify and map these vast empty regions.

Redshift Surveys: Redshift surveys are among the most effective tools for detecting voids. By measuring the redshifts of galaxies, astronomers can determine their relative distances and velocities. In a redshift survey, telescopes collect spectral data from a wide sample of galaxies across the sky, then use this data to construct a 3D map of galaxy distribution. Voids are identified as regions with few or no galaxies in these maps. The Sloan Digital Sky Survey (SDSS) has been instrumental in void detection, producing detailed maps of galaxy distribution over billions of light-years. SDSS data revealed numerous voids, showing them as large-scale structures within the cosmic web. These surveys enabled astronomers to calculate void sizes, shapes, and boundaries, and provided crucial insights into their role in the universe.

Doppler Effect: The Doppler effect plays a critical role in measuring the movement of galaxies and understanding the dynamics around voids. By observing the spectral shifts in galaxy light, astronomers can determine whether galaxies are moving toward or away from us. In regions surrounding voids, galaxies tend to exhibit redshifts as they are "pushed" outward by the expanding void. This outward motion reflects the gravitational influence of denser regions, where galaxies are pulled into filaments and clusters. Doppler measurements provide a way to assess the flow of galaxies and, by extension, to detect and characterize void boundaries where this flow changes significantly.

Gravitational Lensing: Gravitationa lensing is another technique used to map voids, especially when studying the effects of dark matter. In gravitational lensing, the gravitational field of a massive object bends the light of objects behind it, creating distorted or magnified images. While voids lack significant matter, gravitational lensing still provides valuable insights by highlighting regions with contrasting densities. By comparing regions with strong lensing effects to those with minimal lensing, astronomers can better delineate the boundaries of voids and confirm the absence of significant gravitational influence within them. Space-based telescopes, like the Hubble Space Telescope, have been instrumental in gravitational lensing studies, allowing astronomers to observe these effects with high precision.

The Very Large Array (VLA) Contributions

The Very Large Array (VLA) has made substantial contributions to the study and detection of voids, particularly through its capabilities in radio astronomy. By observing radio emissions, particularly the 21-cm emission line of neutral hydrogen (HI), the VLA can detect faint signals in regions where optical light is minimal, making it ideal for studying the low-density gas within voids.

One notable case study involves the detection of faint hydrogen gas in the Bootes Void, one of the largest known voids. Using the VLA, astronomers were able to map the sparse distribution of neutral hydrogen within this region, revealing the low-density gas structure and confirming the absence of significant galaxy clusters within the void. This study helped establish that voids are not entirely empty but do contain traces of neutral hydrogen gas that can be detected through radio observations.

Another key VLA contribution was in mapping the cosmic web's large-scale structure. By observing HI emissions across different parts of the universe, VLA studies have provided insights into the boundaries of voids and the behavior of gas within and around them. These observations support the hypothesis that voids act as a scaffolding framework within the cosmic web, where gas flows from low-density regions into denser filaments, reinforcing the structure of galaxy clusters and superclusters. The VLA's radio wavelength capabilities have been crucial for advancing our understanding of these gas flows, which are difficult to detect in other wavelengths due to the low temperatures and densities involved.

Space-Based Contributions

Space-based telescopes have also played a vital role in void detection by providing high-resolution images and multi-wavelength observations that are not affected by Earth's atmosphere. Two notable contributors are the Hubble Space Telescope (HST) and the James Webb Space Telescope (JWST).

Hubble Space Telescope (HST): The Hubble Space Telescope's deep-field imaging capabilities have allowed astronomers to observe faint galaxies and structures along the boundaries of voids. One major contribution of Hubble to void research was its imaging of the Hubble Deep Field (HDF) and Ultra-Deep Field (HUDF), which provided an unprecedented look at the distribution of galaxies across vast cosmic distances. These images helped map out the edges of nearby voids, showing how galaxies cluster along void boundaries, reinforcing the structure of the cosmic web. Hubble's detailed images of these regions have provided essential data for understanding how galaxies are distributed around voids and how voids fit into the larger cosmic structure.

James Webb Space Telescope (JWST): Recently launched, the JWST offers advanced infrared capabilities that promise to deepen our understanding of voids. By observing in the infrared spectrum, JWST can detect faint, distant galaxies and dust-enshrouded objects near void boundaries. This capability will allow astronomers to study the early universe's voids and explore how these regions evolved over time. For example, JWST is expected to reveal details about gas distribution within voids and around their peripheries, providing critical data on how voids expand and interact with surrounding matter. JWST's sensitivity to faint light sources makes it ideal for detecting any early-stage galaxies that may have formed in or near void regions, offering new insights into the relationship between voids and galaxy formation in the early universe.

In summary, the detection and study of galactic voids have been advanced by both ground-based observatories like the VLA and space-based telescopes like Hubble and JWST. Redshift surveys, Doppler measurements, and gravitational lensing have collectively enabled astronomers to identify voids and map their structures within the cosmic web. By combining data from these diverse methods and instruments, scientists are uncovering the properties and behaviors of voids, which serve as key components in understanding the large-scale evolution of the universe.

Chapter 7

CASE STUDIES OF KNOWN VOIDS

Bootes Void

The Bootes Void is one of the largest and most well-known voids in the universe, spanning approximately 330 million light-years across. Discovered in 1981 by astronomer Robert Kirshner and colleagues, the Bootes Void was initially detected as part of a redshift survey that aimed to map the large-scale distribution of galaxies. The discovery of such a vast, nearly empty region within the cosmic web was unexpected and challenged existing models of galaxy distribution. Within the immense volume of the Bootes Void, only about 60 galaxies have been identified, where, in contrast, a typical region of similar size within a galaxy-rich part of the cosmic web would contain thousands of galaxies.

The Bootes Void is significant for several reasons. First, its sheer size suggests that voids are not merely random gaps in the distribution of galaxies but are structured and may form hierarchies within the cosmic web. The presence of occasional galaxies within the void indicates that voids are not entirely empty but rather contain isolated systems that evolved separately from the denser regions of the universe. Observations of the Bootes Void using the Very Large Array (VLA) have helped detect faint hydrogen emissions within this region, confirming that even in the most sparsely populated areas, matter in the form of neutral hydrogen gas is present. The study of the Bootes Void has provided valuable insights into the expansion of voids and the dynamics of low-density regions, showing how galaxies are drawn toward void boundaries over time, reinforcing the cosmic web structure.

Local Void

The Local Void is a significant empty region near the Milky Way, extending about 150 million light-years. First identified in the 1980s, the Local Void has been closely studied due to its proximity to our galaxy and its apparent influence on the motion of nearby galaxies. Observations indicate that the Local Void may exert a "push" effect on surrounding galaxies, causing them to move away from the void and toward denser regions. This gravitational influence has been observed in the peculiar velocities of galaxies within the Local Group, including our own Milky Way, which appears to be moving away from the void and toward the Virgo Cluster.

The Local Void's influence on local galactic dynamics has been studied using various radio and optical telescopes, with significant contributions from the Sloan Digital Sky Survey (SDSS) and Hubble Space Telescope (HST). These observations revealed that galaxies on the periphery of the Local Void exhibit outward radial velocities, suggesting that void expansion plays a role in shaping galaxy motions even at relatively small scales. Studies using the HST have provided high-resolution images of galaxies near the void, allowing astronomers to observe the effects of isolation on galactic evolution. The Local Void's proximity makes it an ideal subject for studying how voids interact with their surroundings and influence the distribution of nearby matter.

Other Notable Voids

In addition to the Bootes and Local Voids, several other prominent voids have been discovered, each offering unique insights into cosmic structure and evolution:

1. Eridanus Supervoid: The Eridanus Supervoid, located in the direction of the Eridanus constellation, is notable for its association with the Cosmic Microwave Background (CMB) "cold spot," an area of the CMB that is cooler than surrounding regions. This void spans roughly 500 million light-years, making it one of the largest known voids. Some studies suggest that the Eridanus Supervoid may have contributed to the cold spot through the Integrated Sachs-Wolfe effect, in which voids influence CMB photons as they travel through low-density regions. Observations from the Planck satellite have supported this theory, indicating that large voids can have a measurable impact on the CMB.

2. Sculptor Void: The Sculptor Void, located near the Sculptor constellation, spans approximately 200 million light-years. The VLA and SDSS have

contributed extensively to mapping this void, which is part of a complex network of voids and filaments. The Sculptor Void has been used as a case study in understanding the hierarchical structure of voids, as it is connected to nearby smaller voids, suggesting that voids can merge and form larger structures over time. Observing the Sculptor Void has provided insights into how matter is distributed within voids and how void boundaries form and evolve in relation to surrounding filaments.

3. Cepheus Void: The Cepheus Void is a relatively nearby void located within the constellation Cepheus. This void is smaller than the Bootes Void but is notable for the sparse population of galaxies within it, many of which are classified as isolated dwarf galaxies. Observations using the Hubble Space Telescope and radio data from the VLA have revealed that the galaxies within the Cepheus Void have unique evolutionary paths, likely due to their isolation from the gravitational infuence of larger structures. These findings suggest that galaxies within voids may experience slower rates of star formation and reduced interactions with other

galaxies, providing a valuable perspective on how environment influences galactic evolution.

These case studies of notable voids illustrate the diverse characteristics and behaviors of voids across different scales. Each of these voids offers unique insights into the structure of the universe, the distribution of dark matter, and the role of dark energy in cosmic expansion. By studying voids of varying sizes and characteristics, astronomers can better understand the processes that shape the large-scale organization of matter, shedding light on the complex interplay between galaxies, voids, and the cosmic web. The contributions from major telescopes, including the VLA, Hubble, and Planck, continue to advance our knowledge of these mysterious regions, emphasizing the importance of voids in the broader context of cosmic evolution.

Chapter 8

THE INFLUENCE OF VOIDS ON
NEIGHBORING GALAXIES

Gravitational Dynamics Near Voids

The gravitational environment near galactic voids differs significantly from that within dense galaxy clusters. In void regions, gravitational forces are weak due to the low density of matter. However, the gravitational effects of voids on nearby galaxies are substantial. Because voids contain far less mass than surrounding areas, they exert a "push" effect on galaxies along their boundaries. This effect results from the gravitational attraction of denser regions outside the voids, which pulls galaxies away from the lower-density void interiors toward the void walls, where filaments and galaxy clusters form. This gravitational dynamic has been confirmed by tracking galaxy motions using redshift data from large-scale surveys, such as the Sloan Digital Sky Survey (SDSS), and radio observations from the Very Large Array (VLA), which allow astronomers to map these outward flows and measure their velocities.

Galaxies located near the edges of voids, particularly those within filaments and walls that border voids, exhibit unique kinematic behaviors. These galaxies tend to move outward, away from the void centers and toward denser regions. This gravitational "push" effect contributes to a phenomenon known as cosmic flows, in which galaxies are pulled into denser clusters or superclusters. Observational data from the VLA and SDSS have shown that this movement effectively shapes the cosmic web, creating well-defined boundaries between voids and dense structures. These cosmic flows help expand voids over time as galaxies are pulled into nearby denser regions, reinforcing the distribution pattern of matter within the cosmic web and further emptying the interior of voids.

Star Formation in Low-Density Areas

Star formation within or near voids is a rare but notable phenomenon. The lack of dense molecular clouds, which are the primary sites of star formation, means that void interiors are largely inactive in terms of new stellar births. However, some galaxies on the edges of voids or within low-density filaments exhibit signs of star formation. These rare star-forming galaxies provide valuable insights into how stars can form and evolve in low-density environments.

One interesting example includes dwarf galaxies located on void boundaries. Due to their isolation, these galaxies often experience minimal interactions with other galaxies, which can influence star formation processes. Observations from the Hubble Space Telescope and the Spitzer Space Telescope have shown that these dwarf galaxies contain active star-forming regions, although at lower rates than galaxies in denser clusters. The low-density environment around these void-edge galaxies reduces external gravitational influences, allowing them to retain their gas longer, which can lead to sporadic star formation over extended timescales.

Additionally, some studies indicate that isolated galaxies in low-density areas may undergo bursts of star formation when external influences, such as minor gravitational interactions or gas flows, trigger star formation. In such cases, void-edge galaxies can serve as models for understanding how minimal external forces affect the star formation process. The low-density environment appears to delay or slow down typical galaxy evolution processes, resulting in extended star formation periods and unique chemical compositions in these galaxies, as observed in data from the VLA and Hubble.

Void-Induced Galaxy Evolution

The environment within and around voids plays a significant role in shaping the characteristics of galaxies located near these regions. Void galaxies, particularly those in void interiors or near void boundaries, display distinct evolutionary traits compared to galaxies in denser regions. Due to the scarcity of nearby galaxies, gravitational interactions—such as galaxy mergers, which are common in clusters—are rare in void environments. This isolation leads to a slower pace of galaxy evolution, with galaxies in voids often retaining their gas reservoirs for longer periods and experiencing lower rates of star formation.

One prominent theory explaining the unique evolution of void galaxies is the "field galaxy" hypothesis, which suggests that galaxies in isolated regions (such as voids) evolve more slowly and retain simpler structural properties. Many void galaxies are small, low-mass systems, often classified as dwarf galaxies. These galaxies generally have lower metallicities than galaxies in clusters, as they have undergone fewer stellar processes that produce heavier elements. Observations from the VLA support this theory, showing that void galaxies contain more neutral hydrogen gas, indicating that they are less evolved in terms of star formation history and chemical enrichment.

Moreover, the lack of tidal forces in void environments allows galaxies to maintain more stable structures over time. Studies from the James Webb Space Telescope (JWST) are expected to provide further insights into the morphology of void galaxies, potentially confirming that they exhibit fewer signs of interaction-induced features, such as tidal tails or disrupted stellar halos. JWST's infrared capabilities will allow astronomers to study the older stellar populations and gas dynamics in these galaxies, revealing how isolation within a void impacts galaxy morphology and internal star formation patterns over cosmic time.

In conclusion, voids influence neighboring galaxies through unique gravitational dynamics, affect star formation rates in low-density areas, and induce distinct evolutionary paths for galaxies in isolation. Observations from ground-based and space-based observatories, including the VLA, Hubble, and JWST, have greatly expanded our understanding of how these vast empty regions shape galaxy evolution and the structure of the cosmic web.

Chapter 9

FUTURE RESEARCH AND EXPLORATION OF VOIDS

Advances in Telescope Technology

The com ng years hold immense potential for the study of galactic voids, driven largely by advancements in telescope technology. Among the most promising projects is the Square Kilometre Array (SKA), a next-generation radio telescope network currently under construction. The SKA, set to be the world's largest radio telescope, will consist of arrays in South Africa and Australia, with a total collecting area of over one square kilometer. Once operational, the SKA will provide unprecedented sensitivity and resolution, allowing researchers to probe deeper into low-density regions of the universe, including galactic voids.

The SKA's capability to detect faint hydrogen emissions at 21 cm wavelengths will make it particularly effective at studying the gas composition and distribution within voids. With this sensitivity, the SKA could map hydrogen structures within voids, providing insights into their density, temperature, and dark matter interactions. Such detailed mapping could confirm theories regarding the role of voids in cosmic expansion and reveal new aspects of the cosmic web's evolution. The SKA's observations will complement data from the Very Large Array (VLA) and other existing radio telescopes, enhancing our ability to analyze void structures and the behavior of galaxies on void peripheries.

In addition to radio telescopes, new space-based observatories like the James Webb Space Telescope (JWST) and the future Nancy Grace Roman Space Telescope promise to reveal details about voids and their surrounding structures at various wavelengths. JWST, with its infrared capabilities, can observe faint, distant galaxies that may lie near or within void boundaries, providing insights into early galaxy formation and the evolution of voids over cosmic time. The Roman Telescope, set to launch in the mid-2020s, will conduct large-scale surveys with a wide field of view, enabling a broader mapping of cosmic structures and improving our understanding of the distribution and scale of voids.

Role of Machine Learning and AI in Void Research

As telescope technology advances, so does the volume of data collected, presenting new challenges and opportunities for analyzing vast datasets. Machine learning (ML) and artificial intelligence (AI) are becoming essential tools in astronomical research, particularly in the study of large-scale structures like voids. AI algorithms are uniquely suited to detect patterns and anomalies in large datasets, making them invaluable for void identification and characterization.

AI has already been implemented in the Sloan Digital Sky Survey (SDSS) and other large surveys to analyze redshift data and identify low-density regions that may represent voids. Machine learning techniques, such as clustering algorithms, can automatica ly detect void-like structures based on galaxy distribution and density patterns. This automated approach accelerates void detection and reduces human error, allowing researchers to process data more efficiently. The potential of AI in void research will only grow as the SKA, JWST, and other telescopes generate larger and more complex datasets.

Beyond void detection, AI also assists in analyzing the properties of voids, such as their size, shape, and evolution over time. Advanced ML models can simulate the gravitational dynamics within voids, predict how galaxies will behave in low-density environments, and test theories about dark energy and cosmic expansion. For example, AI-driven simulations can predict how voids might evolve under different cosmological models, helping to refine our understanding of dark matter and dark energy. As AI algorithms continue to improve, they will enable more accurate and detailed studies of void characteristics, pushing the boundaries of our understanding of these vast cosmic regions.

Predicted Discoveries

The next generation of telescopes and AI-driven data analysis is expected to lead to significant discoveries about the nature of voids. One anticipated breakthrough is a deeper understanding of dark energy and its influence on void expansion. Since voids are highly sensitive to cosmic expansion, they serve as natural laboratories for studying dark energy. Observations from the SKA and Roman Telescope are likely to yield more accurate measurements of void expansion rates, which can help refine models of dark energy and potentially reveal whether its effects change over time.

Another area of potential discovery is the distribution and behavior of dark matter within voids. By mapping hydrogen gas distribution with high sensitivity, the SKA may reveal subtle dark matter structures that could not be detected with previous telescopes. These findings could help verify or challenge current models of dark matter, contributing to a better understanding of its role in cosmic evolution. Discovering dark matter "halos" or filaments within voids would reshape our understanding of the cosmic web, suggesting that even the emptiest regions of the universe contain complex structures.

Finally, the combination of advanced telescopes and AI may uncover entirely new classes of galaxies within voids. Current observations have identified isolated dwarf galaxies on void edges, but improved detection capabilities could reveal faint or ultra-diffuse galaxies previously undetected. Such discoveries would offer unique insights into how galaxies evolve in isolation and contribute to a more comprehensive view of galaxy formation and survival across different cosmic environments.

In summary, the future of void research holds promise for transformative discoveries about dark matter, dark energy, and the evolution of the cosmic web. With the capabilities of advanced observatories like the SKA, JWST, and the Roman Telescope, combined with the analytical power of AI, astronomers are poised to explore voids in unprecedented detail, uncovering the mysteries of these vast, silent regions of the universe.

Chapter 10

IMPLICATIONS FOR COSMOLOGY AND THE FUTURE OF VOID STUDIES

Voids and the Big Bang

Galactic voids are fundamentally tied to our understanding of the universe's origins, particularly within the framework of the Big Bang theory. According to Big Bang cosmology, the universe began as a hot, dense point approximately 13.8 billion years ago and has been expanding ever since. This expansion led to the formation of large-scale structures, including clusters, filaments, and voids, shaped by initial density fluctuations present in the early universe. These fluctuations, captured in the cosmic microwave background (CMB) radiation and observed by missions like the Planck satellite, show that even at the earliest observable stages, matter was distributed unevenly, leading to areas of high and low density.

Voids are the natural consequence of these fluctuations, as regions with lower initial density evolved into the vast empty spaces between the dense galaxy clusters. According to standard cosmological models, gravity drew matter toward the denser regions, forming galaxies and filaments, while less dense regions became voids. The size and distribution of voids align with predictions from the Big Bang model, as they reflect the underlying physics of cosmic expansion and gravitational clustering. Studies using the Sloan Digital Sky Survey (SDSS) and the Very Large Array (VLA) have mapped void structures in detail, confirming that their distribution corresponds with models of cosmic evolution based on Big Bang cosmology.

Observations of void expansion rates also support the Big Bang model, particularly with regard to dark energy's influence on cosmic acceleration. Voids expand more rapidly than dense regions due to lower gravitational pull, a phenomenon that fits well within the ΛCDM model (Lambda Cold Dark Matter), which includes dark energy as a driver of accelerated expansion. Thus, voids are not only consistent with but also help confirm the current cosmological model, acting as a unique observational test for Big Bang-related predictions on the structure and evolution of the universe.

Theoretical Perspectives on Void Dynamics

While the Big Bang model and the ΛCDM framework are widely accepted, some alternative theories propose different explanations for void dynamics. One such alternative is modified gravity theories, like Modified Newtonian Dynamics (MOND) and f(R) gravity, which suggest that the laws of gravity might change on cosmic scales. These theories predict different behaviors for voids; for instance, they suggest that void expansion rates may vary from those predicted by standard gravity models, with some modified gravity models predicting slower void expansion. Observing voids under these frameworks allows researchers to test the applicability of modified gravity, as deviations from expected void dynamics might indicate a need to revise gravitational theories.

Another speculative concept is the multiverse theory, which suggests that voids might serve as boundaries or "bubbles" between different regions of the universe, potentially representing areas where our universe's physical constants or properties vary from those in adjacent "bubble universes." While direct evidence for this idea remains elusive, some physicists propose that the distinctiveness of voids in their extreme isolation could hold clues about the nature of the cosmos beyond observable space. Voids might, in theory, highlight boundaries of influence where our universe's particular laws of physics fade or transition. Although these ideas are highly theoretical, continued study of void properties could either support or challenge the multiverse hypothesis, especially if voids reveal unexpected patterns inconsistent with current cosmological models.

Voids also provide a testing ground for quantum gravity theories, such as string theory and loop quantum gravity, which seek to unify general relativity with quantum mechanics. Some quantum gravity theories predict that the large-scale structure of the universe might exhibit subtle quantum effects, potentially observable in void dynamics. Future high-precision measurements of void expansion and structure may offer indirect evidence supporting or refuting these theories, giving us new insights into the fundamental nature of space-time.

The Role of Voids in the Cosmic Puzzle

Voids are essential to the "cosmic puzzle" because they embody the effects of fundamental forces— gravity, dark matter, and dark energy—across the largest scales of the universe. By studying voids, cosmologists gain insight into the universe's structure, composition, and ultimate fate. Void dynamics reflect the behavior of dark energy, which drives cosmic expansion; therefore, understanding void expansion helps refine models of dark energy and provides clues about how the universe may evolve in the distant future.

Voids also provide indirect evidence for the distribution of dark matter, as they outline the structure of the cosmic web without containing much visible matter. Observations from the VLA and Chandra X-ray Observatory have shown that void boundaries are shaped by dark matter-dominated regions, reinforcing the idea that dark matter acts as a "scaffolding" for the large-scale structure of the universe. By observing the gravitational influence of dark matter near void edges, scientists can gain a better understanding of dark matter's role in cosmic evolution.

Looking ahead, voids will continue to be key areas of study, particularly as new telescopes and observatories come online. The Square Kilometre Array (SKA) and the Nancy Grace Roman Space Telescope are expected to provide deeper observations of void interiors and boundaries, potentially uncovering new phenomena within these regions. High-resolution maps of voids will further test the ΛCDM model and other cosmological theories, while the role of AI in analyzing complex datasets will accelerate discoveries and refine our understanding of voids' influence on galactic evolution.

In summary, galactic voids are not merely empty spaces but are critical to unraveling the mysteries of cosmology. They align with the Big Bang model, offer testing grounds for alternative theories, and provide essential data for understanding the universe's large-scale structure, its accelerated expansion, and the enigmatic nature of dark energy. Through continued exploration, voids hold the potential to illuminate both the origins and the ultimate fate of the cosmos, making them a central focus in the quest to understand the universe.

BIBLIOGRAPHY

1. Kirshner, R. P., Oemler, A., Schechter, P. L., & Shectman, S. A. (1981). A million cubic megaparsec void in Bootes. Astrophysical Journal Letters, 248, L57-L60.

  - Foundational study on the Bootes Void, revealing one of the first and largest known voids in the universe.

2. Sloan Digital Sky Survey (SDSS) Collaboration. (2000–present). The Sloan Digital Sky Survey. AJ, 120(3), 1579–1587.

  - Comprehensive galaxy survey providing large-scale redshift data and 3D maps of the cosmic web, with a significant role in void identification.

3. Planck Collaboration. (2014). Planck 2013 results. XVI. Cosmological parameters. Astronomy & Astrophysics, 571, A16.

  - Data on the cosmic microwave background (CMB) used to confirm initial density fluctuations that influenced the formation of voids and the cosmic web.

4. Very Large Array (VLA) Observations. (Various years). NRAO Very Large Array observations and data.

 - Detailed radio observations providing essential insights into hydrogen distributions, void boundaries, and void dynamics.

5. Blumenthal, G. R., Faber, S. M., Primack, J. R., & Rees, M. J. (1984). Formation of galaxies and large-scale structure with cold dark matter. Nature, 311(5986), 517-525.

 - Early work on cold dark matter theories explaining the formation of large-scale structures and the role of voids within this framework.

6. Hubble Space Telescope (HST) Observations. (1990–present). Hubble Deep Field and Ultra-Deep Field Surveys.

 - High-resolution optical and near-infrared imaging that maps the distribution of faint galaxies near void boundaries, aiding the study of voids within the cosmic web.

7. Chandra X-ray Observatory Observations. (1999–present). NASA Chandra X-ray Observatory.

   - X-ray data providing insights into the hot gas and dark matter structures surrounding voids and informing studies of gravitational dynamics near void boundaries.

8. Webb, J., & Crittenden, R. (2015). The integrated Sachs-Wolfe effect in voids. Monthly Notices of the Royal Astronomical Society, 453(3), 2514-2526.

   - Analysis of void interactions with the CMB through the Integrated Sachs-Wolfe effect, used to study the impact of voids on the universe's expansion.

9. Kazin, E. A., et al. (2010). The Baryonic Acoustic Oscillations survey and the cosmic void. The Astrophysical Journal, 710(2), 1444.

- Survey linking cosmic voids with baryon acoustic oscillations, contributing to our understanding of void scales and their role in the cosmic web.

10. Square Kilometre Array (SKA) Organization. (Upcoming). The Square Kilometre Array Project: Next-Generation Radio Astronomy.

- Planned radio observations expected to enhance void studies, specifically in hydrogen gas mapping and cosmic web structure.

11. James Webb Space Telescope (JWST) Observations. (2021–present). NASA James Webb Space Telescope.

- Infrared observations aimed at studying faint galaxies in and around voids, advancing our understanding of galaxy evolution in isolated regions.

12. Peebles, P. J. E. (2001). Principles of Physical Cosmology. Princeton University Press.

- Foundational text on cosmology, including theories on void formation, cosmic expansion, and the large-scale structure of the universe.

13. Heath, R. & Bruch, B. (2018). Machine Learning and AI for Astrophysics: Case studies in SDSS void detection. Computational Astrophysics and Cosmology, 5, 6.

- Discusses the application of AI and machine learning in identifying voids from large astronomical datasets, with examples from the SDSS.

14. Rubin Observatory Legacy Survey of Space and Time (LSST). (Expected 2024). Survey overview and goals. The Astronomical Journal.

- Large-scale survey aimed at mapping galaxy distribution and detecting voids with unprecedented detail, furthering our understanding of cosmic voids.

15. Nancy Grace Roman Space Telescope Observations. (Expected mid-2020s). NASA Nancy Grace Roman Space Telescope Mission.

   - Future space observatory designed to conduct wide-field surveys, providing deeper insights into voids and large-scale cosmic structures.

This bibliography serves as a foundation for further study and exploration of voids, with references spanning early discoveries, theoretical frameworks, and observational data from both ground-based and space-based telescopes. The ongoing and future missions of telescopes like the SKA and JWST promise to advance void research significantly, expanding our understanding of these vast and complex cosmic regions.

www.ingramcontent.com/pod-product-compliance
Lightning Source LLC
Chambersburg PA
CBHW070119230526
45472CB00004B/1330